The
HUMMINGBIRDS' GIFT

ALSO BY SY MONTGOMERY

How to Be a Good Creature

Tamed and Untamed (with Elizabeth Marshall Thomas)

The Soul of an Octopus

Birdology

The Good Good Pig

Search for the Golden Moon Bear: Science and Adventure in Pursuit of a New Species

Journey of the Pink Dolphins

Spell of the Tiger

Walking with the Great Apes

The Curious Naturalist

The Wild Out Your Window

Children's

Becoming a Good Creature

Condor Comeback

Inky's Amazing Escape

The Snake Scientist

The Man-Eaters of Sundarbans

Encantado: Pink Dolphin of the Amazon

Search for the Golden Moon Bear: Science and Adventure in the Asian Tropics

The Tarantula Scientist

Quest for the Tree Kangaroo

Saving the Ghost of the Mountain: An Expedition Among Snow Leopards in Mongolia

Kakapo Rescue: Saving the World's Strangest Parrot

Snowball, the Dancing Cockatoo

Temple Grandin: How the Girl Who Loved Cows Embraced Autism and Changed the World

The Tapir Scientist

Chasing Cheetahs

The Octopus Scientists

The Great White Shark Scientist

Amazon Adventure: How Tiny Fishes are Saving the World's Largest Rainforest

The Hyena Scientist

The Magnificent Migration

The HUMMINGBIRDS' GIFT

WONDER, BEAUTY, AND RENEWAL ON WINGS

SY MONTGOMERY

ATRIA BOOKS
New York • London • Toronto • Sydney • New Delhi

An Imprint of Simon & Schuster, Inc.
1230 Avenue of the Americas
New York, NY 10020

An earlier version of this material appeared as a chapter in
Sy Montgomery's book *Birdology* (2010).

First Atria Books hardcover edition May 2021

Interior design by Dana Sloan

Manufactured in the United States of America

1 3 5 7 9 10 8 6 4 2

Library of Congress Cataloging-in-Publication Data is available.

ISBN 978-1-9821-7608-2
ISBN 978-1-9821-7609-9 (ebook)

To mothers everywhere, who understand

The
HUMMINGBIRDS'
GIFT

INTRODUCTION

This is the story of a resurrection.

It's not the world-altering tale of Jesus's rise from the dead after crucifixion. (You already know that one.) Nor has this story the dramatic sweep of the saga of Persephone, daughter of the Greek goddess Demeter, whose yearly escape from the underworld brings us the seasons for planting and harvest—spring, summer, and fall—or the Egyptian myth of Osiris, the murdered king brought back to life by the love of his sister. But it's a story of a miracle, nonetheless.

Granted, it's a small miracle. How small? Not much bigger than two bumblebees—for that was the size of Zuni and Maya, two infant, orphaned hummingbirds, so young we couldn't even tell their species, when I first met them more than ten years ago.

When these two baby birds (who turned out to belong to the

species known as Allen's hummingbird) arrived at rehabilitator Brenda Sherburn La Belle's Fairfax, California home, they were at death's door. Nobody knows what happened to their mother. For mere humans to restore them to the wild required an epic effort. It was my great good fortune that Brenda invited me to help.

Over the course of the weeks I spent helping Brenda, I learned just how demanding and fraught was our task. We were tending to tiny creatures as delicate as froth. In fact, they are bubbles: hummingbirds are made of air. Their tiny bodies are crammed with no fewer than nine air sacs, in addition to their two huge lungs and enormous heart. And yet, in order for them to survive, we had to repeatedly insert into their tiny, bubble-filled bodies a giant, pointy syringe that looked, compared to them, like the top of the Empire State Building.

If we didn't, they would starve. If we overdid it, they could pop.

As it was, I was fearful at first that my touch alone would break them. Everything about a hummingbird is diaphanous. Their delicate bones are exceptionally porous. Their legs are thinner than toothpicks; their feet as flimsy as embroidery thread. Scientists began attaching metal bands to the legs of birds to track and identify them as far back as 1890—but the first bands deemed light and safe enough for hummingbirds' fragile legs were not developed till 1960. And they, of course, were only deployed on adult birds. Our infants were more fragile yet.

Baby hummingbirds require constant, diligent, round-the-clock attention. They tax even their mothers, who may make more than a hundred flights a day to find food for their babies. Even with Brenda's decade of experience, we knew we were no match for a mother hummingbird. We would have to work even harder and longer to make up for that. Saving those tiny babies was a big task.

But the rewards, as I was to learn, were even bigger.

It's been more than a decade since I last saw Brenda. Since then, a lot has happened. Her parents have died, as have her in-laws. She told me it set her thinking: "What am I going to do the rest of my life?" The answer was easy. "I'm going to focus on the two things that make me happy, besides my kids and husband and dog: making sculpture, and my hummingbirds."

She no longer rears orphans—it took her two years to train people to take her place—but she is still deeply connected with these glittering sparks of life. She says she always will be. These days, she still fields calls from around the country, and from as far away as Guatemala. One person phoned her from a third-floor balcony, watching through binoculars to see if a mother hummingbird returned to a nest she feared was abandoned. In another case, the caller found a nest in a terrible neighborhood. No, the area was not afflicted with gangs or drugs. It suffered from a paucity of flowers. Brenda counseled the Samaritan to go out and buy nectar-rich, trumpet-shaped, red blooms and surround the tree with the nest with the plants. Now the mother hummer could tank up on the calories she needed to hunt for the bugs her babies craved.

Brenda's house is still abuzz with hummers. Five nectar feeders set at different corners of the house nourish several dozen hummingbirds of four different species while minimizing fighting. Brenda logs in daily to report her sightings with Cornell University's national Project FeederWatch, to help scientists keep track of bird populations.

Recently, Brenda told me, she and her husband, Russ, bought property in Fort Bidwell, in the far northeastern corner of California at the western edge of the Great Basin. It's high desert, with a strip of rich grassland where ranchers pasture their cattle—a mix of habitats that attracts animals from antelopes to foxes, and birds from sandhill cranes to hummingbirds. They are working to transform the old house into a collective studio, where she and other artists—including many Native American artists—can showcase their work. (They're calling the project Yampa Sculpture Path and

Studio after the Paiute name for an edible tuber that sustained the first peoples, one that still grows wild.) "I want it to be a bridge between art and nature—sculpture *within* nature," Brenda explained, "as well as common ground for artists without any prejudices." And they are turning the land into a paradise for pollinators.

Her love for hummers has spiraled out to now include bees, butterflies, and moths as well. (She particularly loves the sphinx moth, also sometimes called the hummingbird moth. "It's got this funny, coiled tongue," she tells me. "It's really hilarious! It looks like a bent straw—and it hovers!" The moth feeds by night on the same plants the hummingbirds feed on by day.)

On their new land, Brenda and Russ are planting groves and gardens. Pollinator plants will be everywhere: cornflowers, with their soft, fuzzy, blue double blossoms of fringed petals; Buddleias, or butterfly bush, with fragrant clusters of red, yellow, orange, pink, purple, and white petals at the tips of arching branches; columbine, bell-shaped, spurred flowers that dangle and nod with the breeze. She's planting salvia, its tiny, tubular, usually scarlet flowers stacked on tall stalks; penstemon, or beard tongue, sporting towers of tubular flowers in colors ranging from crimson to electric blue. Gardens will be peppered with the shrublike spires of lupines, and dotted with paintbrush, its stalks of linear leaves topped with bright red bracts.

The plantings will not just feed pollinators. "By planting trees and gardens, right down to itty-bitty hummingbird flowers," she says, "we can generate microclimatic ecosystems." Creating little oases of shade and moisture, these carefully chosen communities of plants can importantly counter some of the damage from climate change. A University of Singapore researcher, in a paper in *Global Change Biology*, confirms that even small areas have "extraordinary potential to buffer climate and likely reduce mortality during extreme climate events."

It will be beautiful, too: Brenda envisions visitors walking beneath a trellis covered with orange trumpet flowers, with Buddleias on the side and daisy-shaped bee balm on the bottom. "I'd love to walk through an arbor with hummingbirds and bees and butterflies buzzing all round," she says. Hummers, she says, especially love flowers on trellises: the way they bounce up and down from plant to plant reminds her of music. "It looks like pure joy," she says. "There's something very musical about it, even though they don't sing. Maybe that's how they express their song."

One entire acre will be devoted to a particular pollinator garden that will be planted in a spiral. The idea behind it is, she explains, that here, "you spiral down the path, representing your inner journey—accompanied by bees, butterflies, and hummingbirds."

What better creatures to accompany a person on an inner

journey? In Native American lore, bees symbolize community; Celtic myths tell us bees are spirit messengers from the Otherworld. Around the world, butterflies evoke the promise of metamorphosis. And the sparkling, hovering hummingbirds, because they are impossibly tiny and fast and beautiful, are emblems of irrepressible life.

Today, perhaps more than ever before, we thirst for community; we hanker for transformation; we long to reconnect with the incandescence of life. We need to make those inner journeys. But what if there are no bees or butterflies or hummingbirds to accompany us?

It's a growing possibility. A shocking number of pollinators, including hummingbirds, are in dire danger. Honeybees are suffering from colony collapse disorder. Bumblebee populations are crashing. You are 50 percent less likely to see a bumblebee than you were in 1974. Butterfly populations have decreased, according to one estimate, by 33 percent in the last twenty years. Three billion birds have disappeared from North American skies in the past five decades. Audubon's *Birds and Climate Change Report* warns that half of all birds on our continent are at risk, and it singles out four species of hummingbirds—including the Allen's—projected to lose 90 percent of their small breeding range due to human-induced climate disruption by 2080.

When I spoke with Brenda last fall, smoke was so thick in the air she could not open the windows. She had to wear two masks when she went outside: one for Covid-19, and an N95 for the smoke. Not far from her, 68,000 people evacuated in the wake of the Glass Fire, just one of twenty-eight fires actively burning in California. Ash covered her car, the trees, the flowers, and the hummingbird feeders.

She worried about the hummers. "How can they breathe with all this smoke?" she wondered. "I think because of the reddish dark sky they think it's nighttime. I don't know."

Much of her state had become a hellscape. As ash swirled down from the sky, as the air turned dense, toxic, and orange, the word that came up over and over, from newscasters, firefighters, and ordinary citizens was the same: apocalypse. Now, in the face of the apocalypse, Brenda is making hummingbird ice cubes out of sugar water to keep their food cooler—and planning a spiraling garden for hummingbirds, a garden of resurrection.

The Aztecs believed hummingbirds, who are furious fighters, are reincarnated warriors. They've returned to life with swords as beaks, continuing their battles forever in the sky. Early Spanish visitors to the New World, seeing hummingbirds for the first time (they only live in this half of the globe) called them "resurrection birds." They believed that anything that glittered so brightly had to have been made new each day.

Hummingbirds embody so many opposites that their very existence seems a miracle. They are the lightest birds in the sky—and also, for their size, the fastest. These tiny, fragile birds undertake perilous, long-distance migrations. The rufous hummingbird, often seen in Brenda's yard, flies on gossamer wings from Mexican wintering grounds to nesting areas in Alaska.

And because of—not in spite of—their delicacy, hummingbirds can execute acrobatics that no other bird can approach. Alone among the world's ten thousand avian species, only those in the hummingbird family, *Trochilidae*, can hover in midair. For centuries, nobody knew how they did it. They were considered pure magic.

So while Brenda and I knew that saving the nearly naked, bee-sized orphaned hummingbird babies who had arrived half-dead at her house before us would take a miracle, we didn't despair. We reckoned we already had plenty of evidence of miracles all around us. We just needed one more.

Now, more than a decade after we took on those babies, as we struggle in a world battered by pandemic, human-caused climate disruption, and the politics of rage, I see the miracle yet more clearly. This is the gift that the hummingbirds gave to Brenda and me—a gift we want to pass on to you.

The resurrection of Maya and Zuni matters, of course, because these two little lives mattered. They loved their lives like we love ours, and their joy in life is no less important than our own. (In fact, though unlikely, it is not impossible that Maya and Zuni are still alive: at least one banded hummingbird—an individual of the species known as the broad-tailed—was banded as an adult in 1976 and recaptured and released, alive and healthy, in 1987.) That is gift enough.

But there's more. Their story also holds a larger truth, like a parable. If we, mere humans, could help transform these pathetically vulnerable infants to rulers of the sky, then perhaps our kind can heal our sweet, green, broken world.

An incubator is just a glass and metal box, about the size of a double-wide microwave. Nothing much to look at: it's got a couple of dials and a glass front and makes a soft humming sound as it goes about its task of keeping the temperature inside a constant 85 degrees Fahrenheit and maintaining 45 percent humidity. But on this June day in 2008, I have flown across the country, from Manchester, New Hampshire, to San Francisco, California, ridden a bus across the Golden Gate Bridge to Marin County, and finally caught a ride with my friend Brenda Sherburn La Belle, all for the chance to look in this particular machine. For today this incubator is a jewel box, containing priceless living gems.

The machine sits on a table by the window in the guest bedroom at Brenda's house. I hold my breath as she swings open the glass door. She reaches in and removes a small red plastic utility basket—the kind in which parents store their kids' crayons—and places the basket atop the incubator so I can get a better look.

There, raised an inch above the tissues lining the bottom, resting on a pedestal fashioned from the cardboard core of a toilet paper roll, is a cuplike nest the diameter of a quarter. It's soft as cotton candy, made from puffs of plant down and strands of spider silk and decorated with lichens. Inside, facing opposite directions, with short black bills and eyes tiny as a dressmaker's pins, are two baby hummingbirds. Each is less than an inch and a half long. They are dazzlingly perfect, tiny, and vulnerable.

Together they weigh less than a bigger bird's single flight feather. They are probably eleven and thirteen days old. Already the greenish feathers on their heads hint at the opalescent glow that inspires naturalists to call the hummingbird "a glittering fragment of the rainbow," "a breathing gem," "a magic carbuncle of flaming fire." A week ago, these birds were the size of bumblebees, pink, blind, and naked. After their mother disappeared, they spent a night and a morning alone, starving. It's a marvel they're alive.

"When I got them, I was pretty sure at least one wouldn't make it," Brenda whispers. They were then about three and five days old, with no feathers or eyesight—just hunger and need. "The first three days with them, when they're this small, it's really touch and go," she said. "When you get a baby, you don't really know what it went through till it got to you."

A baby in his nest.

But Brenda knows exactly what the babies need right now: two hundred fruit flies. They're best caught fresh, crushed with a mortar and pestle, then mixed with a special nectar supplemented with vitamins, enzymes, and oils. From dawn to dusk, this food must be delivered into the babies' desperately gaping mouths by syringe, every twenty minutes. Because the food spoils easily, a fresh batch must be concocted several times a day. Brenda is one of a small handful of volunteer wildlife rehabilitators willing and qualified to do so.

I was lucky enough to meet her through a mutual friend, while Brenda was visiting family in New Hampshire the autumn before. Instantly, I liked her. A sculptor and mother of three grown children, she's a five-foot-three powerhouse in dark bangs and a pageboy: her skills range from casting her own bronze to founding an art collaborative for kids to raise funds for conservation. But perhaps even rarer than her energy is her patience. As we sipped tea in my kitchen the afternoon we first met, she answered my questions for more than two hours, speaking in careful, sometimes halting, always thoughtful phrases of the intricacies of hummingbird rehabilitation. The work demands extraordinary precision and commitment. The plights of her charges are often pathetic. Yet, as she spoke, her brown eyes shone with merriment. "The word 'cute' was really invented for a baby hummingbird!" she said. "They are so cute and so fast, so curious and smart—and yet so little is known about them.

"To put a little hummingbird back in the wild," she told me, "might seem like a little thing. But it's a big thing." What is it like to restore these tiny glimmers of birds to the sky? I wondered. Brenda was generous enough to invite me to come to California to the home she shares with her husband, Russ, to observe her work.

Most veterinarians will tell you that orphaned and injured birds are far more difficult to treat than mammals. In birds, often the first symptom of illness is death. Birds are physiologically very different from us. Humans and our fellow mammals are fluid-filled creatures. Early Greek physicians believed all medicine could be based on an understanding of these fluids, which they called humors. But birds, in order to be freed for flight, cannot afford to be loaded down with heavy fluids. Birds are made of air.

Unlike our thick, marrow-filled bones, most birds' bones are hollow. Even their skulls are scaffolded with passageways for air. Their feather shafts are hollow, and the feathers themselves, like strips of Velcro, are interlocking barbules for catching air. Their bodies are filled with air sacs, which originate in, and function, in part, as extensions of the lungs. No fewer than nine of these filmy bladders fill the tiny body of a hummingbird: one pair in the chest cavity; another under each shoulder blade; another pair in the abdomen; one under each wing; and one along the neck.

Hummingbirds are the lightest birds in the sky. Of their roughly 240 species, all confined to the Western Hemisphere, the largest, an Andean "giant," is only eight inches long; the smallest, the bee hummingbird of Cuba, is just over two inches long and weighs a single gram.

Delicacy is the trade-off that hummingbirds have made for their unrivaled powers of flight. Alone among birds, they can hover, fly backward, even fly upside down. For such small birds their speed is astonishing: in his courtship display to impress a female, a male Allen's hummingbird, for instance, can dive out of the sky reaching sixty-one miles per hour, plunging from fifty feet at a rate of more than sixty feet per second—and pulling out of his plunge, he experiences more than nine times the force of gravity. Adjusted for body length, the Allen's is the fastest bird in the world. Diving at 385 body lengths per second, this hummer beats the peregrine falcon's dives at 200 body lengths per second—and even bests the space shuttle as it screams down through the atmosphere at 207 body lengths per second.

Hummingbirds' wings beat at a rate that makes them a blur to human eyes, more than sixty times a second. For centuries, people deemed hummingbird flight pure magic. Until the invention of the stroboscope, scientists could not understand how hummingbirds hover. With a flash duration of one hundred thousandth of a second, the stroboscope finally revealed the motion of wings that had been too fast for other cameras to capture.

Hummingbirds are less flesh than fairies. They are little more than bubbles fringed with iridescent feathers—air wrapped in light. No wonder even experts who are experienced with other birds are intimidated by this fragility. "Their feet are like thread," another rehabilitator, Mary Birney, who lives in Pennsylvania, told me. "Touching them damages their feathers. Yes, they are made of air—air and a humongous heart. That's all they are. It floors me I'm able to work with them."

Hummingbird rehabilitators are unsung heroes. Toiling away with their syringes and tissues, each is a Mother Teresa, a Saint George, a little Dutch boy with his finger in the dike—desperately trying to fend off the hordes of monstrous perils facing these tiniest of all birds. Hawks, roadrunners, crows, jays, squirrels, opossums, raccoons—even dragonflies and praying mantids—eat them. Bass leap from ponds to gulp them whole. Fire ants and yellow jackets sting babies to death in the nest. Flying adults get impaled on the stamens of thistles. They are killed by unseasonable freezes—and by other hummingbirds.

They spar with needlelike bills, but most hummers kill rivals by chasing them away from nectar sources. The losers starve.

They die from infestations of mites. They get blown off course on migration and run out of energy. They fly into spiderwebs while hunting for bugs, or while gathering the silk for nest making. They fall to the ground with their wings bound, mummy-like, in sheets of sticky silk, unable to fly or feed. One woman found such a victim on the floor of her barn, so dirty and lifeless-looking that she kicked it with her shoe before realizing it was not a clod of dirt but a glittering, still living hummingbird, imprisoned in a robe of cobwebs.

The hummingbird's world becomes yet more hazardous with human disturbance, especially the relentless destruction of swamps and woodlands—hummingbirds' best nesting areas, full of nourishing bugs as well as energy-giving flowers. We kill these beautiful birds, too, with our pets. The most common reason for any bird's admittance to a wildlife rehab center is abbreviated on forms as "CBC": Caught by Cat. Hummingbirds also smack into our windows and are hit by our cars. (They are so attracted to red they may kill themselves hitting the glass pane separating them from a red flower in a greenhouse. They will chase red cars. A woman with a scarlet Chevy Suburban reported that immature hummingbirds would probe for nectar in the cracks of its hood.) And because they are all lung and so small, hummingbirds are extremely vulnerable to our pollutants and poisons. Brenda has seen all too many hummingbirds poisoned by common garden pesticides, for which there is no antidote.

Mary Birney, a former pharmacy technician with the University of Pennsylvania, once treated an adult ruby-throated hummingbird who had been trapped in a garage and gotten stuck in spilled polyurethane. The homeowner brought the miserable victim to an animal rescue center. Volunteers used canola oil to remove the sticky goo, but then the bird was oiled, like some poor cormorant after *Exxon Valdez*! A staff veterinarian tried to wash off the oil, but the force of the water coming from the faucet ruptured one of the bird's air sacs. By the time Mary picked up the hummingbird to bring it home, it had swelled up like a toad with subcutaneous emphysema. Mary used a 60 cc syringe to gently wash the bird again—and even this ruptured another air sac. Yet, in the face of all these horrors, Mary observed, "hummingbirds have such an amazing life force." After keeping it for weeks in her home, she released it, healed, to the wild.

Alas, successes with injured hummingbirds are rare. Most medical treatments kill them. Broken wings or legs can't be splinted—a splint would be too heavy (though Brenda, with help from a mentor, once successfully splinted a hummer's toes with a strip of stiff paper). Injections risk puncturing an air sac. And to a bird who weighs as little as a penny, even the slightest overdose of medicine can be lethal. The survival rate for birds admitted to WildCare, the wildlife rehab center with which Brenda works in nearby San Rafael, is impressive: the last year for which stats were available showed that 46 percent were saved (only slightly lower than the 54 percent survival rate for mammals). But that's not representative of sick or injured adult hummers. Almost all of them die.

Sometimes kindhearted people do orphaned and injured hummingbirds more harm than good. One would-be rescuer had fed a baby hummingbird lentil soup and managed to coat its feathers in the bargain. Another gave a hummingbird yogurt and honey (both of which could be lethal). One woman brought in a bird in convulsions. The hummingbird had hit a window; she had tried to treat it with a drug from her own medicine cabinet. Now it was suffering from an overdose of Valium.

Too often, people "rescue" babies prematurely. "People find a hummingbird nest and panic," says Brenda. If the babies are alone, they assume the nest is abandoned. But to feed her young, a mother hummingbird must leave the nest ten to 110 times a day. WildCare staff always tell callers to watch the nest for at least twenty minutes to make sure the babies are really abandoned. "The problem is," Brenda says, "so few people will sit still and just watch something for that long." Invariably their eyes wander from the nest and they miss the mother's lightning-quick return. Some callers won't take the time to watch at all. They'd rather pluck the nest from the tree and take it to WildCare than be twenty minutes late to work.

Orphaned babies do much better than injured adults, though. They have an excellent chance with Brenda, who started working with hummingbirds in 2001 and apprenticed for a year with her mentor, a professional veterinary assistant. And unlike in the East, with its single hummingbird species, the ruby-throat, the San Francisco Bay Area hosts four hummingbird species. Brenda has worked with them all. Black-chins, with their shiny dark heads, white collars, and violet neck bands, migrate through the area. The rufous, with its emerald crown and reddish back spangled with green sequins, passes through on its astonishing trip from Mexico to nest in Alaska—the longest migration of any bird on Earth in terms of body lengths. Anna's hummingbirds, named in honor of Anna Massena, duchess of Rivoli, live year-round in Brenda's garden; the males' throats glow like "roses steeped in liquid fire," as one writer put it. And Allen's hummingbirds, looking very much like the rufous, also nest in Brenda's neighborhood. In fact, the species was named after bird collector Charles Allen, who also lived in the Bay Area. Some hummingbirds here may start nesting as early as February; some may raise second clutches as late as August.

So when Brenda phoned me on a Wednesday in June to say that she had received a pair of orphaned twins, I booked the first ticket available, to fly out that Monday. She was offering me an extraordinary opportunity: the chance to watch these most gossamer of birds grow into masters of the air.

At first, in their exquisite nest, the babies seem as still as stones in a jeweled ring. But suddenly, with shocking speed—if I had blinked I would have missed it—the larger one whirls in the nest to face the opposite direction. Both babies now face south. In a trice, the other baby spins in response. Both snuggle featherless breasts into the nest's soft down lining. Because it's woven with spider silk, the nest stretches to fit new positions precisely. The babies go still again, but their black eyes are open and full of life.

When they came to WildCare, they were nearly dead. "They were not moving or responding," Brenda said. "They were nudies"—without feathers—"and so dehydrated, their eyes, even though they were still closed, seemed to sink into

Baby hummingbirds must be fed every twenty minutes.

Sy and Brenda with orphaned baby hummingbirds.

With stubby orange bills, closed eyes, and pinfeathers, these orphaned infants are little bubbles wrapped in hunger and need.

Days later, the babies' eyes open and their feathers blossom.

Mother hummingbirds weave spider silk into their nests to stretch as babies grow. But separate nests were provided after the larger baby kicked the smaller one out.

A baby in his nest.

Anna's hummingbirds, too, frequent Brenda's backyard.

A male broad-tailed hummingbird. In the mountain forests and meadows of the American West, the metallic trill of its beating wings is a common sound of summer.

The gorgeous sparkling violetear is widespread in the highlands of northern and western South America.

Hummingbirds live only in the western hemisphere, and Ecuador, with its 132 species, is considered a hotspot. Here, a female booted racket-tail, left, and a purple-throated woodstar were photographed feeding from the same flower.

their heads. They looked like little skeletons." WildCare staff put them in a closed oxygen cage; Brenda came to pick them up within an hour. She feared they might not survive the fifteen-minute drive back to her house—particularly the smaller, weaker one.

Most species of hummingbirds usually lay two eggs, which hatch two days apart. The developmental difference between twins can be striking. The larger one is sleeker, fatter, faster, stronger. Its bill is perhaps an eighth of an inch longer. The younger one looks a bit rumpled, with a puff of down still on its head. Both still have a smattering of pinfeathers. Sheathed in a protective wax, the pinfeathers will eventually blossom like flowers, but for now they look like the quills of a miniature porcupine.

The birds grow fast. "Everything's accelerated with hummingbirds," Brenda explains. If the mother has chosen a particularly good nest site, near rich sources of bugs and nectar, her brood can transform from eggs the size of navy beans to flying hummingbirds in as little as two weeks. Offspring from sites farther from food sources might take three times as long to fledge. "It takes us longer, because we are not mother hummingbirds," Brenda explains. "If we do our job well, we can get babies fledged within a month. Every day, you see them grow."

In the week she has known them, they have transformed before her eyes. When they arrived, a novice could not have guessed they were hummingbirds: they were pink blobs, with yellow beaks so stubby they were barely visible. But now they are looking more and more like the astonishing birds they will, with luck, become.

Only yesterday they revealed their species: the rufous color appeared on their backs, identifying them as Allen's hummingbirds. The similarly colored rufous hummers had already flown north. A migratory species that summers only along the coast of California, the Allen's is known for its curiosity and interest in humans. "We're so lucky they're Allen's!" she said. "They're endangered—so every one of them we save—"

As if some inner alarm has rung, Brenda glances at her watch, then checks the timer she has clipped to her waistband. Often, she needs neither: she just knows. "Oh!" she says. "It's almost time."

She bends toward the babies, puffs her cheeks, and gently blows on them. They know this signal well: before their eyes had opened, before the mysterious disaster took their mother away, a soft whoosh of wings as the female landed on the nest promised a meal coming. In response, each gaped its stubby bill to receive her needlelike beak down its tender throat like a sword swallower.

This they do now, with furious enthusiasm. Brenda grabs the 1 cc syringe that has been sitting in a dish next to the incubator. It's fitted with a slender catheter instead of a 24 gauge needle, loaded with fruit-fly-nectar cocktail. The little beaks pop open, and they wave their heads frantically, like an overeager schoolchild waves his arm in the air to attract the attention of the teacher. Brenda thrusts the catheter into first one throat, then the other, pushing the plunger to dispense 0.01 cc into each.

The feeding inspires the older bird to a spectacular feat. He wiggles backward in the nest, jouncing the smaller sibling, nearly executing a headstand. He pushes his tail high above the nest's rim—and, with impressive force, blasts out two turds, each about the size a mouse might leave. They fly more than thirty times the length of his body, past the rim of the nest, past the basket, past the table, and splatter on the window. To my horror, they look like clots of blood. Does the baby have bloody diarrhea? Brenda, calm as still water, is unperturbed.

Russ is blessed with a brother who "has a great compost pile," she explains serenely. "Russ's big job is netting the fruit flies fresh." He does so two or three times a day. One can order frozen fruit flies from suppliers—their usual market is zoos and reptile hobbyists. "But these are bloodier and yummier," Brenda says. "You can tell they like these better. They practically lick their chops! They're such little dinosaurs."

Indeed, the babies do seem to have enjoyed their meal. Why not feed them more?

"Because you can pop them," Brenda says. Miss a feeding by twenty minutes and they can expire. Feed them too much and they can explode.

Brenda resets the timer.

"When I have the babies, we always have to eat something quick for dinner," Brenda explains. Twice while we are preparing the meal (sushi, salad, bread, tortellini with pesto) and twice while we are eating it, the timer goes off. Immediately she goes to the birds to feed them. "Everything," she says, "stops for the hummers."

This sort of schedule makes it difficult to find reliable volunteers to foster baby hummingbirds. "I've had people who said they wanted to—and it's a lot of training," she tells me. "But once they finally get a baby, they realize this is *every day.* After a couple of days, they don't want to get up at five in the morning to get things ready; they don't want to change their schedules at night; they don't want to feed the bird every twenty minutes.

They start slacking off. And you can't do this with a baby hummingbird."

Once, she had an apprentice who was raising an orphan. Brenda dropped by her house to bring her some supplies. The woman wasn't home. "So I waited for her," Brenda remembers. "I assumed she had taken the bird with her. She came driving back and was surprised to see me. I said to her, 'I wanted to give you this stuff for the bird—and where is the bird?' It's in the house—and the poor little thing is almost dead." She took the baby, and under Brenda's care, it survived.

As I watch the twins, I wonder if I could commit to such a schedule, dividing each day into twenty-minute segments, crushing bugs and mixing fresh nectar, day after day. The younger one stretches out a wing and attempts to preen, revealing the membranous skin. Tissue paper seems like armor in comparison. The beaks, too, are extremely fragile, soft and growing. In *The Secret Garden*, Frances Hodgson Burnett wrote of the "immense, tender, terrible, heartbreaking beauty" of eggs. But an eggshell is a fortress compared to the feathers and gossamer skin that shield this minute nestling from all the evils of the world. And yet—now it stands in the nest and tests its wings, whirring with a concentrated ferocity. The wings beat so fast that when we take a picture with Brenda's digital camera, the photo shows a bird with mist for wings. How can someone so fragile be so fearless? How can it summon a resting heart rate of five hundred beats a minute, revving to fifteen hundred a minute when, one day, it flies?

These little bubbles of spunk inspire extraordinary tenderness. One autumn, a ruby-throat, on its lonely, five-hundred-mile migration—a journey across the Gulf of Mexico, which can demand twenty-one hours of nonstop flight—landed, spent, on a drilling platform on the Mississippi coast. It was too exhausted to continue. The oil company dispatched a helicopter to fly it to shore. The hummingbird spent the winter in a gardener's greenhouse, then left, fat and healthy, on its spring migration. Hummingbird rehabbers told me other tales of devotion from ordinary people: folks who find an injured or orphaned hummer are often willing to drive for hours to make it to the rehab center. And now I know for sure: I would do anything—anything—to protect these precious infants who have fallen into our care.

Tonight, as the baby Allen's sleep in the incubator, my nest is a futon just yards away. Their bedroom is my bedroom, and their schedule will become mine: once darkness falls, I won't turn on a light, to avoid disturbing their sleep. I'll carry a flashlight to navigate down the hall to use the bathroom. If I want to read, Brenda offers kindly, I can do so in the living room, but tonight I don't want to. I feel full of my last sight of them, as Brenda replaced their red basket in the incubator after their last feeding, at 8:32 p.m. In their soft nest, they sleep in the shape of a smile: tails curled, backs arched, beaks facing upward toward a sky they will one day own. I almost can't wait for dawn.

5:30 a.m.: In her pink bathrobe, Brenda wakes me with quiet taps on the door. I've woken many times during the night. Fighting the urge to check on the babies, I've been waiting for this moment, when she takes the babies from the incubator to place their basket by the window, so they can wake to natural light. But when Brenda opens the incubator, there is a shocking surprise: the older twin has pushed its smaller sibling out of the nest! I feel sick at the thought of the little bird tossed from its thistledown nest, alone and frightened on the bottom of the basket for who knows how many hours during the night. The baby sits stunned but safe on the tissue-lined floor of the red basket.

"Probably this happens a lot of the time in the wild," Brenda says. Unfortunately, when people find baby birds in this predicament, they often hesitate to replace them in the nest; they're afraid to touch them, for fear a human smell will linger and repel the parents. This is patently untrue. Though birds can detect aromas—and the sense of smell is more important to them than previously thought—most parents greet a baby returned to their nest with visible joy.

Brenda replaces the younger twin in the nest; to my astonished delight, she places the other in my cupped hand. "I'm putting you in a separate nest because you are naughty!" Brenda says jokingly to the older twin. She has a supply of extra hummingbird nests she has saved from previous orphans and goes off to the living room to choose one. Holding the baby, I try to sense the weight and shape of his belly, but I can't. He's as light as a sunbeam, and the skin on my palm feels thick as a shingle. I am a lumbering monster cradling a single breath in my clumsy, oversized hands.

Brenda returns with an Anna's nest for the older twin. It's lined with feathers as well as plant down, and while it, too, is decorated with lichens for camouflage, it has a greener cast to the outside. Brenda sets a small perch, a twig from one of Russ's fruit trees, on either side of the new nest, should the older bird graduate to this new skill. She sets the younger baby in the old nest just inches from the new one so the siblings can see each other and gently places the older baby inside.

The bigger twin immediately spins, shaping the new nest to his body. The younger one, now alone in the original, stretches out, rather like a person in a Jacuzzi: ahhhh! "I've never seen a hummingbird actually lounging in a nest before!" says Brenda. With both babies settled, Brenda and I move on to Phase I of breakfast: grinding coffee and crushing fruit flies.

Our day is ruled by inch-and-a-half-long birds. During the daylight hours, there is no possibility of working on a sculpture, or shopping at the grocery store, or visiting a friend's house, or a workout at the gym. Brenda can't paint, or glaze her clay or ceramic sculptures, or work with pastels. For the weeks before her babies are weaned, she saves stacks of wildlife magazines to read. She sketches ideas for new sculptures. In twenty-minute snippets, she works on Saving the World. (The children's art cooperative she founded, Saving the World One Drawing at a Time, elicits drawings of wildlife from children, which are sold on her website, saveworlddraw.org, to benefit conservation projects the children choose.) Every twenty minutes, a baby hummingbird is hungry. What could be more important than that?

Russ, too, submits to the tyranny of the nestlings. He dashes to the store for the supersized Kleenex that line the hummingbirds' basket, trolls compost piles for fruit flies, picks up takeout for dinner. Everywhere his lanky six-foot figure is followed by four rescue dogs. Foxy, a white Shepherd mix, has lived with the couple for most of her twelve years, but the other three dogs—Scout, a ninety-pound black Shepherd mix, Star, a black Lab mix with a white spot on her chest and a limp, and Buddy, a rambunctious white Lab mix—are recent acquisitions. They came to live with Brenda when Russ was stationed at Fort Leavenworth, Kansas, the previous year. Just months from his sixtieth birthday, the Army Reserve had called him to serve. Within the first two months of his year-long residency, Russ sent home six homeless dogs he rescued from death row. He found homes for the three he and Brenda did not themselves adopt and founded a campaign to "Save the Dogs of Fort Leavenworth." Seven years before we met, he had rescued homeless bees that were living in a cardboard box at the edge of a mall parking lot in San Rafael. When Russ saw them, he went to his parents' house, borrowed his father's bee suit, and brought the bees home. Now they live in a chest of drawers in the backyard.

Russ is retired; Brenda is learning GPS mapping to supplement the couple's income. But for both of them, it's clear that their real work in the world involves more than making a salary. Fostering baby hummingbirds grew out of a volunteer stint Brenda had completed in Belize. After a hurricane had destroyed an Audubon environmental education center, she designed and built a new one. Russ raised funds from the States when the money ran out. When Brenda came home, she visited WildCare, where she had volunteered before, and ran into a rehabilitator who asked her to help with the baby hummingbirds. She was then facing her fiftieth birthday. Just after 9/11 had turned the mood of so many Americans bitter and helpless, she began saving the lives of the tiny creatures that Spaniards called "resurrection birds." The early Spanish visitors to the New World believed that hummingbirds died nightly and revived again. Surely, they must have thought, something that glittered so brightly was made new each day.

This is the gift these baby hummingbirds offer us: a hand in resurrection. Our lives do not stop for them; they begin again. Every twenty minutes, the birds' appetites call us, from our busy to-do lists and fast-paced schedules, back to life.

The nestlings are so minute that Brenda keeps a magnifying glass beside the incubator. Peering through it, we enter a world in miniature. Each of our tiny birds is a landscape for creatures even smaller than they.

Hummingbird babies are so small that Brenda must examine them with a magnifying glass.

At first it seems a speck is stuck on the beak of the smaller bird. But with the magnifying glass, I can see the speck moving. And now I can see another. Brenda knows what these are and assures me they're harmless. These are creatures whose only habitats are the petals of flowers—and the beaks and nostrils of hummingbirds. Unlike other mites, which suck blood, these subsist on flower pollen. But when a flower's pollen runs out, the creature needs a lift to a new blossom. The mite hitches a ride on the beak of a hummingbird, and at the next visit to a flower rushes out to its new home.

"Look—I'll show you," Brenda offers. We grab a red penstemon from a vase in the kitchen—among hummingbirds' favorite blooms. Gently Brenda pushes it toward the face of the little hummer, so that his beak fits inside the flower's tube. I watch through the magnifying glass. When she pulls away the flower, voilà! The mite has left the beak and moved to the petals. This transfer, I later read, is made at a speed comparable to that of a racing cheetah.

Beak mites seldom venture onto the feathers, but they remind Brenda to check both birds for other possible mites. With a moistened Q-tip, she strokes the feathers gently in the direction in which they are growing, then ruffles them in the opposite direction to view the skin beneath. No mites. But the view through the magnifying glass is revelatory. Each tiny feather—the largest of which is scarcely a quarter inch long—is almost a world in itself.

Feathers are among the most complex structural organs found in nature. Nothing of comparable dimension is stronger. They are made of keratin, the same as a human's fingernails, a horse's hooves, and a rhino's horn—but the keratin in feathers, due to a difference in molecular structure, is even tougher.

A typical bird's feathers outweigh its skeleton. Feathers define a bird. By trapping and moving air, feathers protect the bird from cold and wet, and they enable it to fly. But each feather is, itself, largely air, with a stiff central shaft that is light and hollow and attaches, beneath the skin, to a muscle. Like each scale on a reptile, each feather on a bird can be raised or lowered as needed. The shaft divides the feather into two broad vanes on each side, which consist of parallel branches called barbs. At right angles to the barbs are interlocking branchlets called barbules. Hundreds of pairs of tiny barbules on each barb fit together like tiny strips of Velcro and give the feather its weblike quality. The barbules have small hooklike processes called barbicels. And on the barbicels of some feathers are even tinier branchlets, microscopic hamuli, which allow even more air to be trapped in the feather. When birds preen—running their beaks through their feathers, as these babies do after Brenda ruffles them—they are rezipping the Velcro of these minute connections.

Caught in the feathers, air gives birds their warmth and their flight; in hummingbirds, air even gives them their color. Their jewel-like radiance—emerald, ruby, amethyst—comes not from pigment, as in most birds' feathers, but from air. Except for the flight feathers on wings and tail, the top third of hummingbird feathers lack barbicels and hamuli. Instead, they have elliptical structures called platelets (utterly different from the clotting cells in our blood of the same name) filled with microscopic air bubbles. The shape and thickness of the platelet and the amount of air determine the color seen. These air bubbles diffract light into colors that reflect back in a flash of iridescence.

The color on the throat, or gorget, and head of a male is particularly spectacular. The platelets on these feathers are like flat mirrors, and light reflects in only one direction. This is why the gorget of a male ruby-throat, for instance, dazzles in sunlight but may look black in shade. Hummingbirds know this. By carefully adjusting his position in relation to the observer and the sun, a male purposely flashes his colors to intimidate a rival or attract a mate.

Today our babies only hint at the colors they may one day command. Young hummingbirds look like females, whose drabber plumage helps hide the nest in the shadows. If one or both of these babies are male—though we call them both "he" for convenience—we won't be able to tell their sexes before they leave us for Mexico. The gorget color won't develop till spring. But these babies may grow one day to master the very sky.

Such sophisticated strategies seem far in the future. For now, it is thrilling enough to witness ordinary miracles. Like the one we see this morning at 10:15.

For this feeding, Brenda removes the catheter from the syringe. Instead of squirting the food down the bird's throat, she pushes the needleless, food-filled syringe onto the bird's beak. The beak, thin as a straw, easily fits into the opening. The bird's throat flutters and his black eyes seem to widen. What has happened? Brenda answers: "He's just discovered his tongue!"

The appendage is translucent, thin as embroidery thread, and extends and retracts so quickly that if we were not watching the nest with a magnifying glass we might never see it. At a feeder, a hummingbird extrudes and withdraws the tongue thirteen times a second. Many people wrongly think a hummingbird's tongue is hollow like a straw or sticky like flypaper. But hummingbirds do not sip nectar; they lap it. The tongue is forked, like a snake's, with absorbent fringes along the edge of each fork, and grooved down the center to withdraw extra nectar through capillary action. The tongue is so long that, when retracted, it extends back to the rear of the skull and then curls around to lie on top of the skull. (Some woodpeckers, too, have very long tongues—sometimes more than three times the length of the bill—demanding an equally odd storage arrangement when the organ is not in use prying insects from deep holes. In the case of the hairy woodpecker, the tongue forks in the throat, goes below the base of the jaws, wraps behind and then over the top of the woodpecker's skull, and comes to rest inside the bony orbit behind the eyeball.) With this extraordinary appendage, a hummer can drink its own weight in nectar in a single visit to a feeder.

As human babies explore with their mouths and later their hands, hummer babies explore with their tongues. Curiosity rewards a migratory species whose life will depend on finding nectar from hundreds of different plants. The smaller bird picks this moment to rev his wings—first just two seconds, then three, then five—as if giving his sibling a standing ovation for his accomplishment.

From now on, we see the questing tongue often. Through the lengthening bill, the bird extrudes the tongue three times, four times in succession. The baby's feathered face can be remarkably expressive: it is wearing what can only be described as a quizzical look: "What's this? And this? And *this*?"

This slender, silent tongue speaks of hunger for the world outside the incubator. Brenda takes the babies outside. "We're going for a walk," she tells them. Figuratively, of course: we do the walking, carrying the babies in their separate nests in their red basket. The moment we step into the sunshine, they rise from their nests and rev their wings in unison.

We thread along the path through Brenda's four-tiered hummingbird garden. She has planted all the hummers' favorites here: tubular red and yellow columbine, tall hollyhock, crimson fuchsias and salvias, orange lion's mane, dainty coral bells, penstemon, sticky monkey, gooseberry, and currant. The garden is peppered, too, with hanging hummingbird feeders and shallow birdbaths and Brenda's sculptures. Two statues were made from road-killed deer. In the garden, they are resurrected, their skeletons partially clothed in tufa, found stone, and concrete. We bathe the babies in the garden's light and shadow.

We bathe them in sound as well. From the seed feeders on the deck come the euphonious calls of chickadees, the bell-like trill of the dark-eyed juncos, the down-slurred whistle of the titmice, the "ank-ank" of the nuthatches, the "zree" of the house finches, and the coo of doves; from the nectar feeders and flowers, the whirr of hummingbird wings. The babies cock their heads and listen. The entire world is new, and their awakening senses are hungry for all of it.

I feel like a new mother with her child in a stroller. *Here is the sweet, green world!* my heart silently promises the babies. *And one day, all of it will be yours.*

By 6:50 a.m. on Wednesday, our third day together, we're already on the third feeding of the morning. "The little one is really looking like a little Allen's," Brenda observes as she looks at him through the magnifying glass. Almost all the pinfeathers have blossomed into green iridescence on the back, with hints of orange on the head and tail. The bigger one extrudes his long tongue toward this strange big eye, the magnifying glass, wondering what it is.

The glass shows a worrisome development. "I see a lot more mites today," says Brenda. "Especially on the little one."

At 7:14 a.m. we do another beak-to-flower transfer. More than a dozen mites rush off the small one's beak to the flower, but others seem quite content to remain on the baby's head. Both birds are not only preening; now they're scratching themselves with their feet—a lot. Through the glass, we see that there are at least two different kinds of mites on the birds: some are tan, others reddish. Some may be beak mites—but others are clearly not.

More than 25,000 species of mites afflict the world's birds. Some live in the air sacs of canaries, looking like specks of pepper. Others burrow inside the faces and legs of parrots. Chiggers, which also plague humans, beset turkeys and chickens. With piercing mouthparts like those of their relatives, the ticks, the mites feed on the host bird's blood. They itch, wreck sleep, and can cause feather damage as the bird desperately scratches to relieve the itching. Severely infested birds can develop sores, lose weight, and die from anemia.

"That may be why the bigger one kicked the littler one out of the nest," Brenda says. Since mothers reuse the old nest if they raise two broods, nest mites are common in a season's second clutch. Brenda swabs both babies with a moist Q-tip. She picks up five or six mites with each stroke and scrapes them off onto a tissue, which we take to the garbage can in the garage. "These mites can multiply really fast," Brenda says ominously. "I've seen babies totally covered. It can kill them." She falls silent. I can now count more than thirty on the little one, coursing like corpuscles over and under the feathers. He scratches his head with a thread-thin foot and shakes it. They are probably crawling in his ears. It would drive me mad.

"What can we do about this?" I ask Brenda. "Isn't there some insecticide . . . ?"

To Brenda, the idea is almost unthinkable. "Oh! I really don't want to treat him for mites. The stuff could kill him. He's so little . . ."

We try interim measures: While I hold the smaller bird in my hand, Brenda microwaves the original nest. We change the tissue lining the basket. We keep swabbing the babies to remove mites.

But it's a losing battle.

"This is a terrible choice," Brenda says. "The mites can kill them. Or we can kill them, trying to get rid of the mites. They could die from the poison. They could die from the stress. It's so scary!" But it's clear what we have to do.

We put it off as long as we can. We wait till Brenda's new apprentice, Julie Hanft, arrives. A ponytailed mother of a preschool son, she's an eager and competent volunteer who has already worked with WildCare to educate people about coyotes. We wait till the day warms up, so there is less chance the babies will get a chill. Finally, at 10:52, Brenda lets me give the babies one last feeding, while she sets up the kitchen for the treatment. I reset the timer, but they probably won't want to eat afterward.

Brenda covers the kitchen table with paper towels, like a nurse draping the surfaces of an operating room with sterile cloths. She sets out plastic surgical gloves; a clean, empty dog bowl; another clean dog bowl filled with warm water; a pile of Q-tips; a tiny stoppered glass vial of insecticide.

"They won't like it," Brenda announces. "They don't like being out of the nest. Everything is going to be traumatic for them. We'll try to do everything quickly." We steel ourselves for what's ahead.

This is why wildlife rehabilitators seldom name their charges. They'll tell you it's unprofessional. Of course Brenda knows that these babies aren't hers to name, that they aren't pets. But the real reason we haven't named these birds is this: the looming possibility they will die and break our hearts.

Brenda removes the larger one from the nest. Though they have much to learn, nestlings already know one thing: at all costs, stay in the nest. The baby struggles, ferociously clutching the nest's down lining with feet so delicate I fear they'll be pulled from his body. Finally Brenda gets him free. Julie microwaves the nest. When we remove it, a dozen mites lie dead on the paper towel.

Brenda dons surgical gloves, rolls a Q-tip in the mite powder, then strikes the swab on the lip of the vial. Excess powder puffs off like smoke. Brenda places the baby in the empty dog dish on a tissue. She narrates to Julie while she works: "Roll the Q-tip over the bird's feathers, around the neck, up the belly . . . You have to be so careful! This can kill them! . . . Now to the top of the head, now under the beak, and on each side." When she is done, the bird sits in the green dog bowl for a second, innocently stunned by this senseless horror perpetrated by those he trusts. And now for the worst part.

Removing the gloves, Brenda cups the baby in her hand, then dunks him in the bowl of water. He struggles and peeps, as if begging for mercy. "Oh, I'm sorry, I'm so sorry!" Brenda whispers. She takes him from the water, dunks the Q-tip, and wipes the feathers. She dunks the bird again, up to his eyes. The baby looks like his whole body is crying. Wet and bedraggled, the chick seems to have shrunk by a third—a wraith wrapped in wet feathers, a withered leaf plastered to a rain-slicked sidewalk. Brenda hands him to Julie, who ferries him in his nest to the incubator.

The smaller baby is even more pathetic. He struggles weakly when Brenda removes him from the nest. He, too, cries in the water. His tiny chest heaves with fear as Brenda carries him back to the incubator. This is traumatic and exhausting for everybody.

At 11:14, the timer rings. We finished the whole operation in twenty minutes. But neither bird is hungry. "We'll just let them sit quietly and hope they recover," says Brenda. We leave them in peace, setting the timer to check on them and see if we can induce them to eat.

11:47 a.m.: The larger one perches on the edge of the new nest, laboring to breathe, scratching periodically. The smaller one convulsively shakes his head and wings. Both refuse to eat.

12:13 p.m.: The larger bird seems to be settling down. The smaller bird hunches, eyes closed. We can still see the bugs crawling on him. He shakes his head violently, almost convulsively. Brenda reluctantly decides to powder and dip him again. "I am so, so sorry!" she tells him, voice cracking with emotion. "I had one die after doing this," she tells us. "He was really infested, with a dark ring that fell off by the hundreds. He was an Anna's. Oh, I hate this."

12:21 p.m.: After the second immersion, Brenda again offers both birds food; the larger bird accepts a tiny squirt and then turns away, as if the food tasted bad or made him feel sick. The other refuses.

12:45 p.m.: "Want some? Want some?" Brenda blows on the babies and waves the syringe. The larger one gapes enthusiastically and swallows his first real meal in an hour and forty-five minutes. "Yeah, you look good!" she says to him. "He's just a little older, and his tolerance . . . plus he didn't have as many bugs." He rises and beats his wings, looking strong as a crowing cock. The food has revived him. This bird is going to make it.

But the smaller bird doesn't want to eat. Brenda manages to get him to accept a tiny squirt. He swallows weakly and then droops in the nest, his eyes slits. "Please, please, please be okay," Brenda whispers.

"I think we need to be out of the room," she says.

We move to the dining room, a welcoming space with a fireplace and a view through sliding doors of the deck and seed and nectar feeders ablur with the wings of purple finches, titmice, orioles, towhees, grosbeaks, and hummingbirds. Inside, shelves and cabinets display treasures from the couple's life: rocks, fossils, antlers, crystals of red hematite and quartz, petrified wood, whalebone from the beach, an ancient anchor stone from an Indian canoe, an Audubon print of the water ouzel (also called the water thrush), and several of Brenda's smaller sculptures. They are abstract, with rough edges and holes, and look almost like they were found, not made. The holes let in light, which seems to change the sculpture; it's as if the holes let in time and movement as well. The forms remind me of kachina dolls, the movement of a walking ape, of beings part human, part animal. I ask her about her art.

Growing up in the mill town of Plymouth, New Hampshire, Brenda read an issue of *Life* on Matisse, and another on Picasso, and it changed her life. She dropped out of high school at sixteen and moved to Boston—"to find a mentor to apprentice to and be a sculptor." I looked at her incredulously. "That's what you did in the Renaissance," she said. "I didn't know any better." She found one immediately. "On a side street, near Harvard Square, beneath a bookstore, there he was, working in his studio. He was perfect! I went up two or three nights and pounded on the windows and he just yelled at me to go away. One night I yelled, 'I want to make sculpture and I want to be your apprentice!' And he yelled back, 'I teach at New England School of Art and Design—if you want to work with me, enroll!' I was devastated."

College seemed unlikely. Brenda's mother was the only one in the family to graduate from high school. Her father left school to join the navy in World War II and worked as a tool grinder and machine specialist at a knitting factory for twenty years—till it was bought by outsiders who closed it down. He lost his pension.

But Brenda did go to New England School of Art and Design and worked with the basement sculptor, Robin Binning. She funded her tuition with an inheritance left by her grandfather's sister—for a year and a half. Then came her first marriage, two daughters and a son, as well as the son of her husband's sister, whom the couple adopted when he was three—when Brenda was just about to go back to school again. But always she has found time to make her sculptures.

"Sculpture is about process, and this process is often fragmented. In a way, our lives are like that. A lot of people have trouble with transitions, with discontinuity. But this is what makes us grow. It's mysterious. You shift to a new platform and see things from a different perspective. Art does that. Wildlife does that. Wildlife and art reveal these transitions and demand we experience them. They—"

The timer sounds.

And at last, the smaller baby is hungry. Both birds eat with gusto, eyes bright. The littler one preens a wing feather with his bill—a balletic, careful motion, not a tortured one. "Very good!" says Brenda. And at last, the triumph: he rises in the nest and, clutching its softness with his feet, revs his wings. Brenda and I hug and wipe our eyes.

"You know that kind of awestruck, timeless feeling you get when you look at a great work of art?" Brenda says. "That sense of wonder, that sense of connection to something great and mysterious? It's the same feeling looking at a baby hummingbird."

Each is just a speck—a firefly, a flash, a brilliant atom. Yet each is an infinite mystery.

"It's a layer of this world we know very little about," Brenda says.

It's now that I propose to break the rule. "Let's name them."

Brenda concedes. Our interest in these babies is far more than professional. Frankly, we love them—not just because they are hummingbirds, but because they are *these* hummingbirds, distinct individuals. Besides, we're tired of calling them "the bigger one" and "the smaller one." Brenda asks me for suggestions.

I've been thinking about this. I suggest we name the older, stronger bird Aztec. The Aztecs admired hummingbirds so deeply that they adorned statues of Montezuma with their feathers. Aztecs believed that hummingbirds, forever chasing one another with swordlike beaks, were resurrected warriors, returned to life so they could continue their battles in the sky. Their word for the god of war, Huitzilopochtli, combines the Aztec words for "hummingbird" and "sorcerer who spits fire."

"Too violent!" says Brenda. "Something else."

Every culture to encounter hummingbirds has tried to name their magic. Ancient Mexicans called them *huitzitzil* and *ourbiri*—"rays of the sun" and "tresses of the day star." In the Dominican Republic, people call them *suma flor*—"buzzing flower." The Portuguese called them *beija-flor,* or "flower kisser." Even the scientists succumbed to hummingbirds' intoxicating mysteries: they classified them in an order called Apodiformes, which means "without feet"—for it was believed (incorrectly) for many years that a hummingbird had no need for feet. It was thought that no hummingbird ever perched, accounting in part for its sun-washed brilliance: as the comte de Buffon, Georges-Louis Leclerc, wrote in his 1775 *Histoire naturelle*, "The emerald, the ruby, and the topaz glitter in its garb, which is never soiled with the dust of the earth."

But for names we'll speak many times a day, I suggest two others. For the larger bird, Maya: the Mayans believed that hummingbirds were made from scraps left over from other birds and that their brilliant colors were a parting gift from the Sun God. For the smaller one, Zuni: like the Hopi and Pima, they believed that hummingbirds bring rain—which this part of California hasn't seen since January. Apt names, I feel, for both are named for blessings. And before long, Maya and Zuni will be embarking on the arduous journey to seize the greatest of all birds' blessings: the blessing of flight.

Bird flight is a confluence of miracles: Scales evolved into feathers. Marrow gave way to air. Jaws turned to horny, lightweight beaks bereft of teeth. (This is why many birds swallow stones: to grind their food since they can't chew it.) Hands grew into wingtips. Arms became airfoils. "The evolution of flight has honed avian anatomy into an extreme and remarkable adaptive configuration," anthropologist Pat Shipman writes in her wonderful book *Taking Wing: Archaeopteryx and the Evolution of Bird Flight.* But while most birds are made to fly, and the urge to fly is instinctual, flight itself must be carefully and painstakingly learned—a task of impressive complexity.

Consider the three basic methods of bird flight. The simplest, gliding flight, demands exquisite balance and judgment. Gliding flight exploits passive lift to counteract the pull of gravity. Vultures, hawks, and eagles ride the currents within thermals, rising columns of warm air. Albatrosses and petrels exploit different layers of windspeed above waves. Birds can glide for hours, expending very little energy.

Flapping flight—the way most birds fly—is more demanding still. Achieved by flexing wings at joints in wrist, elbow, and shoulder, it is powered by extraordinarily strong breast muscles. The wings move forward in a downward arc, propelling the bird forward and up. It is similar to the oarsman's power stroke or the action of a swimmer doing the butterfly. Movement then flows into the upward stroke, a recovery stroke, to start the process anew.

And finally, there is hovering, unique to hummingbirds. No other bird really hovers—kites, storm petrels, kestrels, and kingfishers appear to do so, but only hummingbirds can sustain this method of flying for more than a few moments. Instead of flapping the wings up and down, the wings move forward and backward in a figure eight. During the forward and back strokes, the wings make two turns of nearly 180 degrees. The upstroke as well as downstroke require enormous strength; every stroke is a power stroke. Like insects and helicopters, hummingbirds can fly backward by slanting the angle of the wings; they can fly upside down by spreading the tail to lead the body into a backward somersault. Hovering becomes so natural to a hummingbird that a mother who wants to turn in her nest does it by lifting straight up into the air, twirling, then coming back down. A hummer can stay suspended in the air for up to an hour.

Hummingbirds are specially equipped to perform these feats. In most birds, 15 to 25 percent of the body is given over to flying muscles. In a hummingbird's body, flight muscles account for 35 percent. An enormous heart constitutes up to 2.5 percent of its body weight—the largest per body weight of all vertebrates. At rest, the hummingbird pumps blood at a rate fifteen times as fast as that of a resting ostrich, and that blood is exceptionally rich in oxygen-carrying hemoglobin. "There can be no doubt it reigns supreme over all the other birds of the world," Esther Quesada Tyrrell writes in *Hummingbirds: Their Life and Behavior*, "and truly deserves to be called the champion of flight."

While Maya and Zuni are still revving wings in the nest, I will fly home on a jet plane. By the time I return, they will be flying on their own—and nearly ready to fly free.

Years of experience have taught hummingbird rehabilitators how to ready baby birds to become champions of flight. No human

The sword-billed hummingbird takes the long hummingbird beak to extremes. It's reported to be the only species with a bill longer than its body.

Another Ecuadorian beauty, the violet-tailed sylph.

An aptly named tourmaline sun angel, native to Colombia and Ecuador.

Hummingbirds reign supreme for agility in flight, as these buff-tailed coronets, native to cloud forests in Colombia, Ecuador, and Venezuela, demonstrate.

These glittering gems can fight viciously. Coronets may lock feet like little eagles, spin to the ground, and continue fighting.

Glittering like sequins, hummingbirds' iridescence comes from special structures, not pigments, in their feathers. This is a crowned wood nymph, found in northern South America.

Though tiny and fragile, hummingbirds, like no other birds, own the sky.

can teach a hummingbird to fly, of course. Rather, Brenda thinks of herself at this stage as "a little gym instructor." The best she can do is provide the fledglings, at the right time, with the properly equipped gymnasium.

When an orphaned baby first leaves the nest and starts perching, it graduates to a larger basket, where it learns to fly from one perch to another. When that task is mastered, the bird moves to a laundry basket, where it begins to hover. All the baskets are encased in nylon mesh to prevent escape—and also to hold in the live fruit flies Brenda and Russ release there. Catching bugs is a further incentive to hone depth perception, strengthen muscles, and develop beak-eye and wing-tail coordination.

During my absence, Brenda emails me news of the babies' milestones:

July 1: Maya moving to #2 basket and we will introduce fruit flies. FUN. Zuni just fledged—not self-feeding. Zuni peeps for me to feed him.

July 3: Zuni is FINALLY self-feeding (yesterday) but still in #1 basket. Maya just moved to a #3 basket. I will move Zuni to a #2 basket tomorrow . . . transitions seem to be more challenging for Zuni.

July 6: Maya hovering and catching fruit flies in the #3 basket. He is the perfect bird, right on track. Zuni is in the #2 basket, flying better, but still gapes in the a.m. I keep the two side by side outside by day and move them in at night. Moving Maya to small wire cage today.

The small wire cage is about two and a half feet by two feet by two feet. The bottom of the cage is covered with potted blooms, all shades of coral, pink, and crimson—penstemons, fuchsia, *Bravoa geminiflora.* The top half of the cage is uncrowded, with space for unobstructed flying. If you could hear the buzzing in there, you might conclude that this is a cage in which crazy people are keeping a bumblebee.

The bumblebee, I find when I return, is Zuni. I hear him long before I see him, and the sound, so loud and deep, makes me laugh. Its pitch is the mark of an amateur, Brenda tells me: "He hovers, but he's not good at it yet," she explains. "When he gets it, he'll make just a little buzz." Russ rushes outside to get him more bugs, so Zuni can demonstrate fruit-fly catching. The little bird zigzags after the flies, seizing them in a tweezer-like bill that has grown at least a quarter inch since I last saw him. Zuni's flights last only seconds and sound like a propeller plane—but to me, they are miraculous.

Next we visit the porch and the large white flight cage on the deck. About five feet tall, the cage once housed a pet parrot. But now it's been rechristened the Hummingbird Hotel. Furnished by Brenda's artist's eye, it's like a lovely garden inside, the floor carpeted with nine potted plants, the walls adorned with syringes, perches, feeders, and water for bathing. The cage is a beautiful setting for the stunning gem inside it: glimmering, sleek, green Maya. He zooms up and down, flitting from one bloom to another, hovering at exactly the right angle to extract nectar from the throat of each flower. His wings are a blur. It looks as if he is riding in a small tornado. "Isn't he great going up and down with his tail?" Brenda asks. "This is one of the things he had to perfect in this cage."

Maya has learned all he can learn from this largest of Brenda's cages. Tomorrow will be a momentous day: he will fly free and begin his life as a wild Allen's hummingbird.

In the morning, Brenda turns Maya's big cage so it faces the garden. We move Zuni's cage from the dining room table to next to the Hotel. When you move a cage, the hummingbird in it flies—as if Maya and Zuni both realize this is the proper way a hummer moves from place to place, rather than being carried.

The day warms. Zuni buzzes, rocking from side to side a bit as he flies. Maya flits from flower to flower, bathes, preens, then softly buzzes from one side of the cage to the other, as if pacing. Maya seems ready for the release. But I wonder: is Brenda? She has done this many times. This is, after all, the point of all those weeks of every-twenty-minute feedings, of curtailing her artwork and social life, of tears and prayers. How does she feel? "I feel confident I did everything I can," she says. "He can fly up and down. He can recognize the feeder, he knows many plants. This is what I want. I really want them to be wild birds." But every time she releases a hummingbird, her heart, too, flies out that cage door.

She summons Russ. "It's time."

Brenda draws aside the curtain of the screen and ties it back for Maya. He hovers, perches, then flies in zigzags from top to bottom as if the cage has grown too small for him. He seems eager for the wider world.

"Watch—there he goes!" cries Russ. But Maya is still buzzing inside the cage. He can't see the way out. He sits on the lower perch in the middle of the cage, as if considering his options.

"Sometimes," Russ tells me, "they fly out fast and—*bip!*—they're gone." Other times, they just sit awhile, understandably hesitant to leave the protection of the cage. And sometimes, Russ said, they fly out, and then back in, before they finally take off.

Maya feeds from the flowers at the bottom of the cage. He flits to his feeder. And now—at 11:01—he darts out of the cage and dives down a slope into a bay tree below the house, a male Anna's in hot pursuit. It's a flight easily in excess of sixty feet—twenty times farther than he has ever flown before.

Zuni peeps piteously in his flight cage. In the wake of his sibling's achievement, "he wants his food, he wants his mother," says Brenda. Though there are feeders he can reach easily in the flight cage, she feeds him by hand. "I'm worried about him. I've been worried about him ever since we got him."

We scan the yard with binoculars. Released Anna's always vanish, but Allen's are different. They're likely to stick around for a few days, so we hope we'll see Maya again. We want to document his first days of freedom. Brenda leaves the Hotel door open. She wants to test her hypothesis that newly released Allen's welcome the chance to roost there in safe familiarity for their first few nights of freedom.

But Maya has seemingly vanished.

11:30 a.m.: We check for Maya on a circuit every thirty minutes. We first peer from the garage in the direction of the Hummingbird Hotel. We next check the feeder by the studio. Two male Anna's shoot by like flaming comets. A third hovers at the door of the vacant Hotel, eyeing the flowers and feeders inside. Finally we walk through the hummingbird garden. The feeders and flowers buzz with Anna's of both sexes. But no Maya.

12:00: Two Anna's zip through the garden, fighting. A newly fledged Anna's tries to feed by the studio but is chased away, again and again, by adult males, flashing their rose gorgets. The fledgling huddles, fluffy and forlorn, in the branches of an oak. But where is Maya?

"You wonder," says Brenda. "It's a big world . . ."

"He'll come back," says Russ gently.

"Sometimes he would flag and sputter like a little helicopter in trouble . . ."

"He was flying fine," Russ answers. "He'll be fine."

12:30 p.m.: We watch with binoculars from Brenda's studio. The world is crowded with hummingbirds. Anna's zoom everywhere: at the corner feeder, at the lily of the Nile, at the red salvias. "I don't see anything," says Brenda. "Where is my baby?"

What scares us most is we have not seen him feeding. To survive, a hummingbird must consume the greatest amount of food by body weight of any vertebrate. A film Brenda loaned me claimed that a person as active as a hummingbird would need 155,000 calories a day—and the human's body temperature would rise to 700 degrees Fahrenheit and ignite! To fuel the furious pace of its life—even resting, it breathes 250 times a minute, and its heart pounds at five hundred beats per minute—a hummer must daily visit fifteen hundred flowers and eat six hundred to seven hundred insects. If the nectar alone were converted to its human equivalent, that would be fifteen gallons a day.

Food is so precious to hummingbirds that they defend "their" flowers and feeders against all comers. They sometimes even chase away hawks and crows. Their main rivals, of course, for the food are other hummingbirds, and hummers' belligerence toward their fellows is legendary. Russ recalled the title of one article about them headlined "So Little. So Pretty. So Mean." In *Hummingbirds: Their Life and Behavior,* Esther Tyrrell and her photographer husband, Robert, document fighting hummingbirds trying to stab each other's eyes out with their bills. The couple has seen a hummer grab the rival's bill in flight and, locked in midair, the two birds fall to the ground together—then rise to continue the battle. A male hummingbird may spend one minute in fifteen lapping from a favored feeder—and the other fourteen defending it. But the way hummers usually kill rivals is bloodless. They simply chase the rival bird from food until it runs out of energy. It can enter a state called torpor in which the body temperature, normally more than 105 degrees Fahrenheit, falls to below 70. (And recently a University of Pretoria researcher reported that the internal temperature of one species of Andean hummingbird, the black metaltail, drops to under 38 degrees Fahrenheit! Some species of these high-altitude hummers can sustain torpor over twelve hours, suggesting these birds are capable of hibernation.) Torpor can also be caused by chill, and it can take an hour for a hummingbird to rouse from it; during this time, the most slow-moving predator—even a possum—can take a hummingbird. But if a predator doesn't get it, starvation will. A torpid hummingbird may perch next to a nectar-laden flower but be too weak to summon the energy to drink from it.

Is Maya now torpid, frozen with fear and exhaustion? Normally a fledgling hummingbird would call its mother to feed it. But we can neither hear nor see our baby. If he's in trouble, we are impotent to help. I feel a growing lump in my throat. Please please please, my heart pounds. Just a glimpse. Just one more glimpse.

It's as if Brenda can hear my thoughts. "But then, the more

you don't see them, the better," she says. "We really do want them to leave."

1:00 p.m.: A female Anna's hovers, lapping from the purple globular blossoms of a lily of the Nile. A male flashes his gorget as he feeds from a potted fuchsia. A baby perches high in the madrone tree, cleaning its bill on a branch. But it's an Anna's, not our Maya.

1:30 p.m.: Only yellow jackets at the feeder by the studio. An Anna's perches on the lowest branches of the oak. Maya is nowhere to be seen.

1:40 p.m.: Was that him? While Brenda is in the kitchen, I think I see him trying to drink from the studio feeder. But the youngster is chased immediately by a male Anna's with a flashing magenta gorget. A baby Anna's eyes the studio feeder, decides on bugs instead, but is chased away anyway. Another male flashes his deep rose gorget below from the laurel. Was I imagining Maya?

1:53 p.m.: No—it *was* him! As I walk from the studio toward the garage, I find him perched on the wire fence surrounding the deck. He sips from a potted fuchsia, then perches on the wire behind the flowers, trying to hide from the Anna's—but they chase him off instantly, careening after him like bullets. He buzzes my face, hovering in front of me, as if he is asking me for help. I rush in to get Brenda.

When we come out, he's perched on the wire again, fluffed up and peeping like his life depended on it—and it might. Maya weaves his head back and forth like a desperate child, as if pleading: Help me!

"He looks really stressed out," says Brenda. Just as a mother hummer would do, Brenda rushes to feed him. She holds up one of the extra feeders she's put out along the fence. Maya hovers before it—drinking, at last, unmolested. He perches, then hovers for a second drink. This helps. He stops peeping. He preens. He cleans his bill with his foot, then wipes his beak on the wire.

"It's a rough, tough world out there," says Brenda. "All those Anna's attacking! These first couple days are so critical. Between the other hummers and all the predators out there . . . but he has to do it . . ."

Brenda brings more nectar feeders and sets them along the fence. Attached to the fence with a wad of putty, a syringe can become a hummingbird feeder in a jiffy. Brenda has an almost endless supply. The idea is to overwhelm the Anna's with abundance so little Maya gets a chance to feed. People who set out feeders and plant wisely for hummers know this well. It's possible to have as many as forty hummingbirds in your yard at once—even though every bird thinks every feeder and flower belongs to him alone.

Despite the abundance, the Anna's continue to chase. But Maya is often able to steal a drink before he flies off. He's learning, and he's clever. At 2:48, he flies to the studio feeder and is chased away by the Anna's. We're impressed with the sophistication of his escape: he maneuvers his tail to whip and loop down toward the laurel below and around the poison oak to safety. At 3:50 he perches on the wire and is challenged bill to bill by a hovering male Anna's—and holds his ground. Sometimes he uses one of us as a human shield. At 4:10, he reappears on the wire near the fuchsia when he is dive-bombed by an Anna's. Both fly toward Brenda, who is watering the hummingbird garden. She offers the spray to the Anna's, who takes a bath—and now, with feathers soaked, will need to dry and preen, giving Maya a chance at the feeders.

"You're getting it, baby!" Brenda cries. We hug and high-five. From the kitchen, Brenda brings sparkling water in long-stemmed glasses. "A toast to our little baby!" she proposes.

But it could be too early to celebrate. At 5:20, Maya

reappears, feeding on fuchsia flowers by the garage door. In a flash he vanishes, shooting over the roof with a speed and purpose we had never before seen. No other Anna's are in sight. In fact, even the seed feeders are curiously vacant. Then—BAM! Something hard and heavy hits the wood railing of the deck, just yards from where Maya had perched. It's the steely talons of a sharp-shinned hawk.

Twenty-nine minutes later, Maya returns. He perches on an unused bird cage near some potted fuchsias. As two Anna's fly overhead, he appears to flinch, but doesn't flee. When a turkey vulture swoops close to the porch, I watch carefully to see what Maya will do. Humans often confuse vultures with hawks, as both are big birds who soar. Many field guides class vultures with raptors, even though hawks are hunters and vultures eat carrion. But Maya doesn't make the same mistake. He watches the vulture with interest but does not shrink from it or fly away.

How does a five-week-old hummingbird know the difference between a deadly hawk and a harmless vulture? The knowledge could be innate. Or it could be the result of careful observation. Hummingbirds are curious and astute observers. Backyard hummers quickly learn to recognize individual humans and approach people who feed them—even if they are not carrying food. (Sometimes they'll hover at windows to attract a particular person to come out and refill an empty feeder.) In *Hummingbird Gardens*, authors Nancy Newfield and Barbara Nielsen write about a Costa's hummingbird whose nest fell apart in an exhibit at the Arizona-Sonora Desert Museum. While an assistant constructed an artificial nest, keeper Karen Krebbs held the five-day-old nestlings. Though wary of the assistant, the mother bird clearly recognized and trusted the keeper—so much that she perched on the woman's hand as she fed her babies. Hummingbirds remember what they see. Many folks who feed hummers report that if they're slow to get their feeders out in the spring, hummers show up and hover right where the feeder used to hang the year before.

Maya's quick flight from the hawk, and his worldly nonchalance with the vulture, give me some reassurance. Maya is a competent little bird. But he's still a baby. And night is coming, with its cool temperatures, hunting owls, and roaming cats. I'm glad Brenda's leaving the door open to the Hummingbird Hotel.

The release cage is a safe haven. Once, two wild Anna's took refuge there during a winter storm. One chose a perch near the top; the other chose the farthest possible spot away from him. Normally hummingbirds can't stand the sight of another hummer who's not a potential mate or its own young. They eyed each other warily but spent the night together nonetheless.

At 7:20 p.m. we find Maya perched serenely on a stem in the release cage. Through the kitchen window, we watch him till dark, until he is just a silhouette that dissolves into the soft, moonlit night.

Having spent the night, happy and comfortable, inside the release cage, by 6:30 a.m. Maya is again a blur: flying in and out of the Hotel, sipping from feeders inside and out, inspecting flowers in the garden. Zuni, too, spent the night outdoors—his first—in the small wire cage right next to the Hotel. "He seems very happy," says Brenda sunnily. "I think they like feeling the cool air."

Today Brenda plans to move Zuni into the Hummingbird Hotel. Since he hatched two days after Maya, that would put Zuni just about on schedule for a release three days from now. Typically birds spend three days perfecting their hovering in the Hotel before release. But Zuni is not a typical hummingbird. "He may panic. Every time we've moved him up a cage, he's panicked," Brenda says.

Once the day has warmed, Brenda captures Zuni in her cupped hands while I hold aside the screening by the door to the Hummingbird Hotel. Zuni perches on an upper right stem. He surveys his new surroundings like a first-time visitor to a great cathedral: he seems awed as he moves his head, bird-jerkily, in every direction above him. But within ten seconds, he hovers to a fuchsia in a top-level vase. Russ releases some bugs in the cage and Zuni flits about snapping them out of the air.

It's a delight watching both babies expand their world. Midmorning, Maya hovers in front of one of Brenda's statues—an armless torso, about three feet high. There seems little hope of nectar here—it's not red, a color many flowers sport specifically to attract hummingbird pollinators, as it's a color bees don't see. The statue is white concrete. There is nothing flowery about it. Maya simply seems curious. He spends nearly a minute examining it—more time than many people invest in any given piece of art at a gallery. Zuni, meanwhile, is clearly practicing his flights and hovers. In the large cage he can cover nearly three times the airspace of the smaller one. He zooms up, down, back, and forth. The stems of flowers bounce when he lands on them. It reminds me of a child playing on a trampoline.

Hummingbirds do seem to play, and to enjoy it. Authors Newfield and Nielsen recount how one Anna's flitted through the spray of a fountain in a California garden: "One day she discovered that she could ride the stream, a solid jet of water about three quarters of an inch thick. Flying at right angles, she alighted on the jet, as though it were a branch, and permitted it to carry her forward. Over and over she did this, apparently enjoying the stunt. She seemed to be playing rather than bathing." A hummer will also bathe on leaves slick with dew or rain. It will slide down the leaf, moistening its breast and shaking its feathers while still in motion. It must feel lovely.

The following morning, we can see how much Zuni's hover has improved. An adult Allen's hovers with an upright posture, holding the tail high with little movement; when moving up and down, it flicks the tail. Up to now, Zuni's posture has been hunched, his tail wavering. "He's getting it!" Brenda cries.

But when we check on him again at 8 a.m., the little bird who was doing so well is now a pathetic, peeping puffball.

He issues a peep every second, one after another. Each sounds like it could be his last. When Brenda approaches, he gapes as if he wants her to feed him. But when she holds the syringe to his beak, he won't stop gaping long enough to drink. Finally he inserts the tip of his beak into the feeder and we see his throat flutter as he flicks his tongue.

What has gone wrong? We back away from the cage and

watch from the kitchen window. And now the problem is clear: Anna's are challenging him from outside the cage. They buzz by, hovering in front of the flowers inside, eyeing them jealously. They discovered the place was full of flowers and feeders while the cage door was open yesterday. The males are desperate for food: though the species is resident in the Bay Area year-round, many individuals retreat from the northern edge of their breeding area in the fall. The males are first to leave, and they must increase their body weight by half for the trip. "It's such an intense time for them," says Brenda, "and there are so many baby Anna's!" To Zuni, it must feel like his house is surrounded by thugs, peering through his windows, eyeing his stuff.

Zuni is right to be afraid. "An Anna's will kill an Allen's," Brenda says softly. Anna's will occasionally stab with their bills but more often they kick their opponents midair as they hover. They taunt each other with a tail gesture that mimics this action. "Instead of giving him the finger, they're giving him the tail," I say. One brash Anna's even dares to perch on the cage bars and stick his swordlike bill inside. Zuni dissolves into peeping, like a child bursting into tears.

"If he's like this inside the cage . . . ," Brenda worries. "He's got plenty of bugs, and more flowers than he'll ever need . . ."

"What can we do?" I ask.

"We might have to move him back to the other wire cage," Brenda says.

"Oh no! That's worse than being left back!"

Brenda thinks for a moment. "He's safe—but he doesn't know it," she says. She slides a glass panel into one side of the cage, to repel the Anna's thrusting bills, and covers another side with a blanket. At least Zuni won't have to look at frightening Anna's bullying him from all sides. She also moves the cage. It's currently positioned, she realizes, directly in the flow of traffic between two of her most popular hummingbird feeders—the one at the kitchen window and the one outside her studio. Just a few yards away, Zuni's cage won't be blocking traffic anymore. And finally, she decides we should go run an errand in town. Perhaps if he can't see us, he won't focus on calling for his mother.

But when we return, Zuni is so despondent he won't even fly to the syringe when Brenda holds it up to him. He's like a tot who has cried for so long he's forgotten what he's crying about—and now is crying because he's exhausted from crying. Brenda has raised three children and more than eighty baby hummingbirds. She knows what to do. Patiently, she holds the syringe to his beak while he perches. Finally he summons the wit to drink.

The food revives him, but soon he's peeping again. Brenda spritzes him with water. He's forced to preen, giving him something to do other than perch frozen in terror. He fluffs, runs his beak through his feathers, wipes his beak on the perch. Dry again, he zooms around the cage as if nothing had happened, but thirty minutes later, he's puffed and peeping again.

"You know how kids get cranky and nothing can console them?" Brenda muses. "I think the Anna's just messed up Zuni's whole day. He's got plenty of food. He's flying well. He can do it—maybe he'll be all right tomorrow."

Zuni improves each day. On Saturday, an Anna's sticks its bill into the Hotel right in Zuni's face—and Zuni holds his ground. When the Anna's leaves, Zuni hovers for two seconds—an aerial victory dance?—then perches on a high stem. Nary a peep. Now he understands the cage is a safe barrier.

Maya buzzes by from time to time, but not as often. He's enlarging his territory. Brenda spots him over in the neighbor's yard. "He was flying so fast you would never see him if he didn't stop!" she tells me. While Maya is drinking from Brenda's studio feeder, a juvenile Anna's zooms in. Like a targeted missile, Maya rockets after him, chasing him from the feeder and over the roof.

Sunday dawns cool and windy. Brenda had hoped to release Zuni today, but the wind is strong enough to lift a tablecloth off the picnic table on the deck. And to our horror, Zuni's right wing is fluffed out at the shoulder. When he flies, he lists to the right. Did he fly up the side of the cage and hit a perch? Did he crash into something while frightened by an Anna's? Was he buffeted by a gust of wind? Whatever it was, it happened while he was safe in the cage. Brenda is exasperated. "This bird has been one problem after another!" she says. Today's problem seems to be a single, twisted feather. Brenda spritzes Zuni with a mister, hoping he'll set the feather right when he preens. Zuni, always an enthusiastic bather, sets about the task rather cheerfully. He's not puffy or peeping; he doesn't seem fearful or in pain. The twisted feather isn't bothering him as much as the setback is bothering us. "He'll be releasable," Brenda promises. "I just can't take the chance to release him today."

But we will celebrate anyway. Brenda has arranged for me to do a reading at a local bookstore, and afterward she and Russ are hosting a hummingbird garden party. Russ's parents and brothers, neighbors, and friends from WildCare are invited; a friend of mine is driving all the way up from Fresno, 200 miles away. Also on the guest list are two people we have never met:

the couple who first found Zuni and Maya. We've always wondered about them.

The original rescuers are the last guests to arrive, after most of the others have left. Michelle Earnhardt, thirty-nine, pretty and petite, has long blond hair and is wearing white pants, a long blue scarf, and a leather jacket. Her heavyset husband, Matt, thirty-seven, towers over her five-foot frame like a big bear in shorts and a T-shirt. She's a hairdresser; he's a contractor. They wanted to get here earlier, Michelle explains, but the drive from Novato, normally half an hour away, was clogged with traffic. It's clear that coming here was very important to both of them.

Michelle tells us how they found the twins. They had just returned from their wedding and honeymoon in Hawaii when they first saw a hummingbird frequenting the passionfruit vine on the trellised entryway to the side yard. They were delighted. It was just six feet from the front door. "We saw a hummingbird there all the time," said Michelle. "And then, one afternoon, it wasn't there."

Hours passed, and though the couple watched and waited, still they didn't see the hummingbird. Both were nearly sick with worry. They both love animals. Before they moved to Novato, they had rented a house in Mill Valley that came with a single goldfish who lived in a barrel. What was he eating? they wondered. He must be starving! So they named him Starvin' Garvin and fed him garden worms. When they moved to Novato, they left Garvin behind—after all, he didn't belong to them. But that first night, Matt couldn't sleep for worrying about the fish. He drove back to Mill Valley in the middle of the night and returned with Garvin and his barrel.

To Michelle, hummingbirds were especially important. She had once gone to a sort of fortune-teller, a shaman, who told her she had a totem animal, a creature who was constantly with her, guiding her, even though she couldn't see it. Her totem was a hummingbird, the shaman said, and it stayed right in front of her forehead. Michelle was not surprised to learn this, she said, for hummingbirds have appeared to her and guided her many times at important junctures in her life.

When she was a teen, her mother and later her friends had taken to calling Michelle a hummingbird, since she was so petite and quick. Michelle had spent part of her childhood in Iowa, and even though the family now lived close to one another in California, at one point Michelle had contemplated moving back. Her mother was thinking about this when one day the song "Hummingbird" came on the radio, with its lyrics "Don't fly away." Immediately afterward, a hummingbird hovered in front of her face. Michelle's mother phoned her daughter: "You can't move. I got a sign." Michelle is glad she listened. If she had moved, she never would have met Matt.

A hummingbird helped her right before her wedding, too, she told us. Just weeks before the wedding date, she, Matt, and ten other people flying to Hawaii for the ceremony lost their tickets due to airline problems. "I was so upset," she said. And then one cool morning, Matt found a hummingbird on the back porch, unmoving. He picked it up and it flew away. "And I thought," said Michelle, "if a hummingbird can stay still that long, I can, too." Suddenly, she felt at peace.

So Michelle and Matt were determined to try to find the missing mother hummingbird and help if she was in trouble. She and Matt leaned a big ladder up against the side of the house so he could get a better view. That's when he spotted the nest, partially hidden in the passionfruit vine. He parted the leaves. Two pink, naked babies popped their heads up, gaping voicelessly for food.

Michelle called WildCare, but the rehab center was closed. The couple didn't know what to do.

That night before they went to bed, Michelle climbed the ladder to check on the babies. They weren't moving. She wept all night.

In the morning, before Matt left for work, he climbed the ladder to check the nest one last time. He brushed aside the leaves, and the larger baby popped open its mouth! The other baby seemed barely alive. But there was hope.

It was 7 a.m. WildCare wouldn't open till nine. So Matt drove the twenty miles to his contracting business in Tiburon, organized his work crews for his absence, then drove back to Novato. He ascended the ladder and clipped the nest free of the passionfruit vine. He placed the nest in a shoebox and, with the babies periodically gaping weakly in the passenger seat of his pickup, drove the fifteen miles to WildCare.

"Want to see them?" Brenda asks. The five of us—Brenda and Russ, Matt and Michelle, and me—head out to the deck. It feels like a family reunion of sorts, for all of us had a hand in Zuni's and Maya's survival. Maya zips by like a comet. "That was one of them," says Brenda. Now they approach Zuni in his cage for a longer look. "Oh!" Michelle whispers, and touches Matt's strong arm with her fingertips. When she last saw this little bird, he was pink, naked, all but dead. Now he is a feathered sequin, hovering before a flower.

"He gets himself in trouble," Brenda tells Michelle. She speaks about Zuni with the fondness of a teacher talking to a mother about a favorite but difficult child. "Somehow he got off-kilter today. But he's okay now. His chances are really good."

In three days, Zuni will be released; two days later, both he and Maya will vanish, headed on the fall migration.

We don't know that yet as we stand before the tiny bird in the release cage. But already, we can imagine him flying to Mexico.

RESOURCES FOR THOSE WHO LOVE HUMMINGBIRDS

The Hummingbird Society

Provides hummingbird information and endangered species protection.

http://www.hummingbirdsociety.org/index.php

Operation Ruby Throat

Offers information about attracting and studying hummingbirds, feeding, banding, and student projects.

http://www.rubythroat.org/

The Hummingbird Monitoring Network

Combines scientific research with community involvement to create projects and services that benefit hummingbirds.

https://hummonnet.org/

WildCare

This is the urban wildlife hospital with which Brenda has worked for a decade.

https://www.discoverwildcare.org/

Yampa Sculpture Path & Studio

Brenda's new nonprofit cultivates a place for enjoying sculpture within nature while creating an oasis for all pollinators, including bees, butterflies, moths, and especially hummingbirds.

https://yampapath.com

Brenda's art website, https://www.saveworlddraw.org, also includes a page about her orphaned hummingbirds:

https://saveworlddraw.org/hummingbirds/

If you find an injured or orphaned hummingbird
please visit: https://www.hummingbird-guide.com
/hummingbird-rescue-care.html

ABOUT THE PHOTOGRAPHER

Originally trained as an oil painter, Tianne Strombeck now creates photographic portraits of nature to promote conservation and wildlife education. Her painting background has given her an instinctive understanding of color and composition. She works to understand her subjects, to capture their essence, and to show how they interact with their environment and each other. Her work has appeared in numerous books, including Sy Montgomery's *The Soul of an Octopus* and *Condor Comeback*. To see additional examples of her work, from hummingbirds to jaguars, visit her galleries at https://www.tianimal.com.